QUANTUM MECHANICS

FOR BEGINNERS

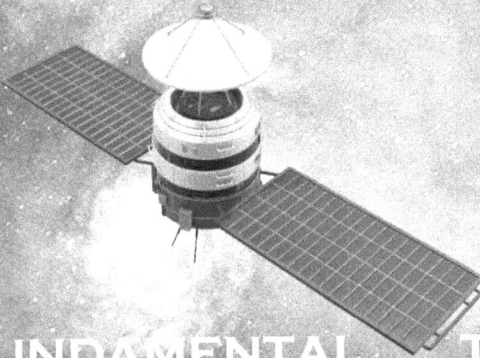

FUNDAMENTAL THEORIES OF
QUANTUM MECHANICS
AND HIDDEN SECRETS
OF THE UNIVERSE
MADE EASY

ALEXANDER SCHLOTTERBECK

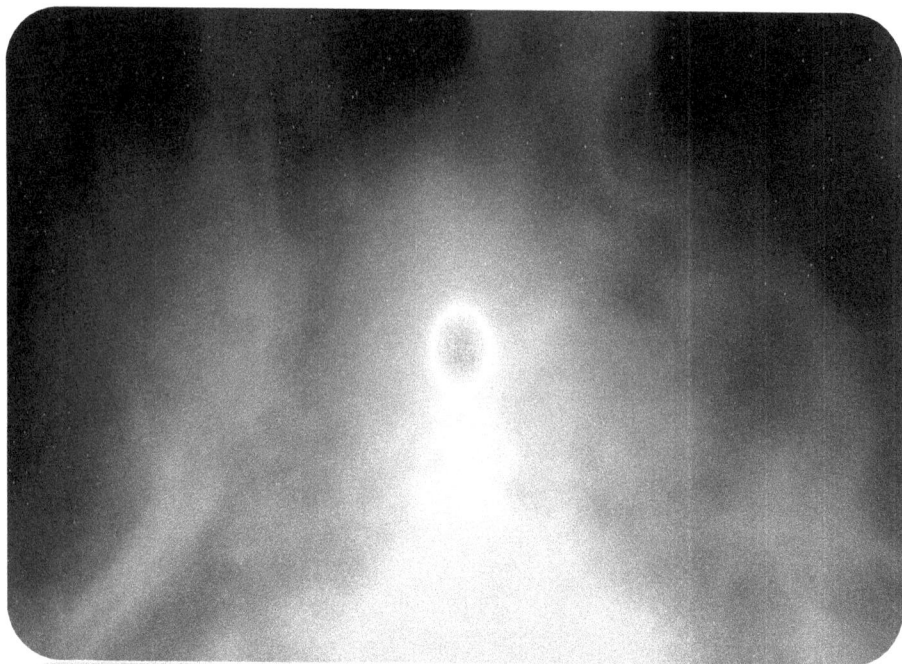

TABLE OF CONTENT

considered an illegal act irrespective of if it is done electronically or in print.

This extends to creating a secondary or tertiary copy of the work or a recorded copy and is only allowed with the express written consent from the Publisher.

<u>ALL ADDICTIONAL RIGHT RESERVED.</u>

The information in the following pages is broadly considered a truthful and accurate account of facts and as such, any inattention, use, or misuse of the information in question by the reader will render any resulting actions solely under their purview.

There are no scenarios in which the publisher or the original author of this work can be in any fashion deemed liable for any hardship or damages that may befall them after undertaking information described herein.

Additionally, the information in the following pages is intended only for informational purposes and should thus be thought of as universal.

As befitting its nature, it is presented without assurance regarding its prolonged validity or interim quality.

Trademarks that are mentioned are done without written consent and can in no way be considered an endorsement from the trademark holder.

INTRODUCTION

In the Maxwell era, physicists had already understood that static electricity is created whenever they rub, for example, a piece of amber with a rabbit fur. They had also noticed that a compass needle moves every time a magnet is placed nearby. Given the very different nature of these effects, these two phenomena were considered independent and unrelated.

However, at about the same time, some critical observations made it clear that electricity and magnetism can be connected. Maxwell derived a series of four simple equations, which showed that electricity and magnetism were only two sides of the same coin. The two phenomena have always been connected by something called an electromagnetic field.
Just as one gravitational field allows any mass to drag on another, Maxwell's electromagnetic field allows any positive charge to repel positive charges and attract negative charges.
Maxwell showed that a flow of moving electrical charges produces an electromagnetic field that could move a compass needle. He proceeded on to say that if those moving charges were to increase in speed or change direction, they would produce an electromagnetic wave that would travel in space.
This wave is a disturbance in the electromagnetic field itself.
Maxwell's classical electrodynamics was very powerful. He explained almost every electrical or magnetic phenomenon known at the time.

For example, it could successfully explain why colors emerged from Newton's prism and why Young's double slits formed a diffraction pattern.

Physicists and engineers still use it today to define many electrical and magnetic phenomena with extreme correctness.

It could also be used to calculate the rate at which electromagnetic waves should travel through space. Maxwell claimed that these waves move at the same speed that physicists had been considered for light rays.

When Maxwell's theory was confirmed several decades, there was little doubt that light was indeed a wave phenomenon.

ELECTROMAGNETIC SPECTRA

Today we know that visible light is not the only type of electromagnetic wave out there.

The radio waves collected by the mobile phone and the microwave are waves that adapt to a broad electromagnetic spectrum.

The only difference between these different types of waves is the speed with which they oscillate: this quantity is called frequency and is represented by the symbol f.

According to classical physics, the electromagnetic spectrum is continuous, and every frequency is allowed. It is also feasible to measure the "length" of electromagnetic waves.

The distance from the wave's crests is known as the wavelength (represented with the Greek symbol λ).

When it comes to visible light, red light has the longest wavelength (lowest frequency), while the shortest purple (highest frequency).

The last quantity needed to describe electromagnetic waves is the speed with which they travel.

The swiftness of light is denoted by the symbol c. This is a constant value that never changes.

It is a universal speed limit since nothing can travel faster than c. Mathematically, the three quantities are connected by the equation $c = \lambda f$.

Since all electromagnetic waves travel at the same speed, longer wavelengths have lower frequencies, and shorter wavelengths have higher ones.

Most light sources, like the sun, actually emit light that spans a range of frequencies.

Physicists also use particular light sources that emit pure light of a single frequency or monochromatic light.

Compared to the human-sized water waves that we observed earlier when we were on the boat, the light waves are much shorter.

The wavelength of the orange light emitted by a lamppost is approximately 60 millionths of a centimeter (0.00006 cm).

It is this "smallness" of the light waves that made Newton curious.

Light waves are noticeably short compared to the size of our mirror, for example.

This means that when the light bounces or passes through it, the deviation from the movement in a straight line is imperceptible.

Therefore, Newton's geometric optics work well for almost all everyday applications.

It no longer works only when light interacts with microscopic objects, such as Young's thin double slits.

If we travel along the electromagnetic spectrum towards gradually longer wavelengths, we will arrive at infrared radiation.

Infrared radiation is not visible to the naked eye, but tools such as night vision goggles can easily detect it. They work by sensing the thermal radiation (heat) emitted by the objects they see.

We stressed that Maxwell's classical electrodynamics could explain almost all the electromagnetic phenomena that we observe every day.

Spectra emitted by heated solids and excited gases, however, are exceptions.

A Nod to Thermodynamics

Although he will be remembered forever as the father of electromagnetism, one of Maxwell's most famous theories had nothing to do with this topic.

In 1873, he turned to the British Association for the Advancement of Science to discuss "molecules." However, he referred more generally to the concept that gases are composed of small particles that move intensely.

He claimed that the air in the classroom was full of molecules traveling in all directions at speeds of around 17 miles per minute. Maxwell and his contemporaries understood that the air temperature and pressure around them were directly proportional to the speed of gas particles.

There are approximately 1×10^{23} particles in the volume of a beach ball. Since the velocity of these will vary somewhat over a specific range, it is more accurate to say that the ambient temperature and pressure are determined by the average velocity of all those particles.

The general relationship between particle velocity, temperature, and pressure is termed thermodynamics. It can be classified as the third and final pillar of classical physics.

As highlighted by Maxwell's important lesson, its center is the small particles that make up the air.

THE FATHER OF QUANTUM THEORY

All objects emit electromagnetic radiation, which is called heat radiation. But we only see them when the objects are extremely hot.

Because then they also emit visible light. Like glowing iron or our sun. Of course, physicists were looking for a formula that would correctly describe the emission of electromagnetic radiation. But it just did not work out. Then, in 1900, the German physicist Max Planck (1858- 1947) took a courageous step.
The emission of electromagnetic radiation means the emission of energy.
According to the Maxwell equations, this energy release should take place continuously.
"Continuously" means that any value is possible for the energy output. Max Planck now assumed that the energy output could only take place in multiples of energy packets, in steps that led him to the correct formula.

To the energy packets, Planck said, "quanta."
Therefore, the year 1900 is regarded as the year of birth of quantum theory.
Important: Only the emission (and the absorption) of the electromagnetic radiation should occur in the form of quanta. Planck didn't assume that it was composed

of quanta. Because that would mean that it would have a particle character. However, like all other physicists of his time, he was persuaded that electromagnetic radiation comprised absolutely of waves. Young's double-slit experiment has revealed it, and the Maxwell equations have established it.
In 1905 an interloper named Albert Einstein was much bolder. He took a glance at the photoelectric outcome. The methods that electrons can

be knocked out of metals by irradiation with light. According to classical physics, the electrons' energy knocked out should depend on the intensity of the light. Strangely enough, this is not the case. The energy of the electrons does not depend on the intensity but the frequency of the light. Einstein could explain that. For this back again to the quanta of Max Planck. The energy of each quantum depends on the frequency of the electromagnetic radiation. The higher the frequency, the greater the energy of the quantum. Einstein now assumed, in contrast to Planck, that the electromagnetic radiation itself consists of quanta. The interaction of a single quantum with a single electron on the metal surface causes this electron to be knocked out. The quantum releases its energy to the electron. Therefore, the energy of the electrons knocked out depends on the frequency of the incident light. However, the skepticism was great at first. Because electromagnetic radiation would then have both a wave and a particle character, but another experiment also showed its particle character. This experiment was conducted with X-rays and electrons carried out by the American physicist Arthur Compton (1892 - 1962) in 1923. As already mentioned, X-rays are also electromagnetic radiation, but they have a much higher frequency than visible light. Therefore, the quanta of X-rays are very energetic. That's why they can colonize the particular form. But that makes them so threatening. Compton was adept at viewing that X-rays and electrons act relatedly to billiard balls when they meet. This again showed the bit character of the electromagnetic radiation. So their twofold nature, the alleged "wave-particle dualism," was lastly acknowledged. By the way, it was Compton who introduced the term "photons" for the quanta of electromagnetic radiation. What are photons? That is still unclear today.

Under no circumstances should they be imagined as small spheres moving forward at the speed of light. Because the photons are not located in space, so they are never at a certain place. Here is a citation from Albert Einstein. Although it dates back to 1951, it also applies to today's situation: "Fifty years of hard thinking have not brought me any closer to the answer to the question "What are light quanta? Today, every Tom, Dick, and Harry imagines they know. But they're wrong."

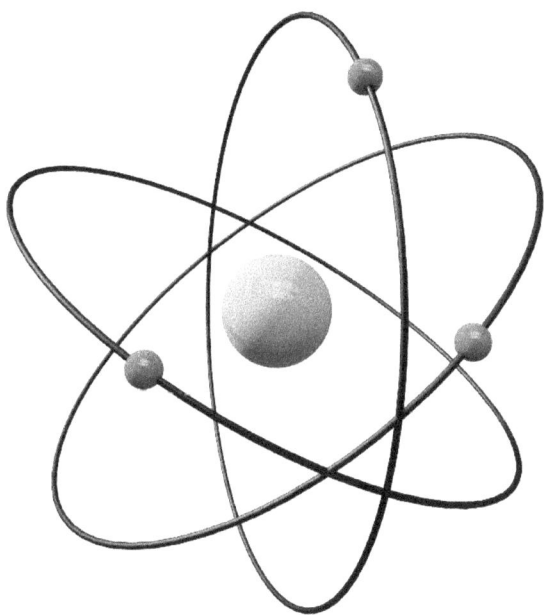

THE BOHR ATOMIC MODEL

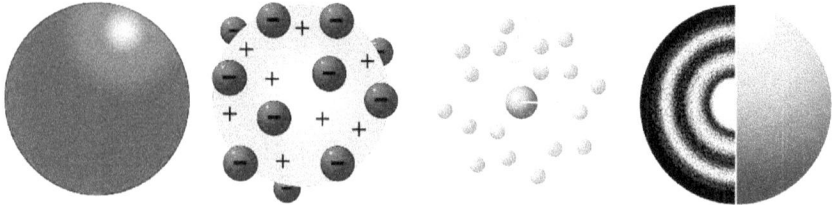

We take atoms for granted. Their existence was still controversial until the beginning of the 20th century. But already in the 5th century BC, the ancient Greeks, especially Leukipp and his pupil Democritus, spoke of atoms. They thought the matter was made up of tiny, indivisible units. They called these atoms (ancient Greek "átomos" = indivisible). In his miracle year 1905, Albert Einstein not only presented the special theory of relativity and solved the mystery of the photoelectric effect, but he was also able to explain the Brownian motion. In 1827 the Scottish botanist and physician Robert Brown (1773 - 1858) discovered that dust particles only visible under the microscope make jerky movements in the water. Einstein was able to explain this by the fact that much smaller particles, which are not visible even under the microscope, collide in huge numbers with the dust particles and that this is subject to random fluctuations. The latter leads to jerky movements. The invisible particles must be molecules. Therefore, the explanation of the Brownian movement was regarded as their validation and thus also as the validation of the atoms.

In 1897, the British physicist Joseph John Thomson (1856 - 1940) discovered electrons as a component of atoms and developed the first atomic model, the so-called raisin cake model. Therefore, the atoms consist of an evenly distributed positively charged mass in which the negatively charged electrons are embedded like raisins in a cake batter. This was falsified in 1910 by the New Zealand physicist Ernest Rutherford (1871 - 1937). With his experiments at the University of Manchester, he was able to show that the atoms are almost empty. They involve a small, positively charged nucleus. Around him are the electrons.
They should rotate nearby the nucleus as the planets rotate near the sun. Another form of movement was inconceivable at that time. That led physics into a deep crisis. Because the electrons have an electrical charge, and a circular motion causes them to release energy in electromagnetic radiation. Therefore, the electrons should fall into the nucleus. Hence the deep crisis because there should be no atoms at all.

In 1913 a young colleague of Ernest Rutherford, the Danish physicist Niels Bohr (1885 - 1962), tried to explain atoms' stability. He transferred the idea of quanta to the orbits of electrons in atoms. This means that there are no random orbits around the nucleus for the electrons, but that only certain orbits are allowed. Each has a certain energy. Bohr assumed that these permitted orbits were stable because the electrons on them do not emit electromagnetic radiation. Without, however, being able to explain why this should be the case.
Nevertheless, his atomic model was initially quite successful because it could explain the so-called Balmer formula. It has been identified for some time that atoms only absorb light at certain frequencies. They are called spectral lines. In 1885, the Swiss

mathematician and physicist Johann Jakob Balmer (1825 - 1898) found a formula with which the spectral lines' frequencies could be described correctly. But he couldn't explain them. Bohr then succeeded with his atomic model, at least for the hydrogen atom. This is because photons can be excited by photons, causing them to jump on orbits with higher energy. This is the famous quantum leap, the smallest possible leap ever. Since only certain orbits are allowed in the Bohr atomic model, the energy and the frequency of the exciting photons must correspond exactly to the energy difference between the initial orbit and the excited orbit. This explained the Balmer formula. But Bohr's atomic model quickly reached its limits because it only worked for the hydrogen atom. The German physicist Arnold Sommerfeld (1868 - 1951) expanded it. However, it still represented a rather unconvincing mixture of classical physics and quantum aspects. Besides, it still could not explain why certain orbits of the electrons should be stable.

Sommerfeld had a young assistant, Werner Heisenberg (1901 - 1976), who, in his doctoral thesis, dealt with the Bohr atom model extended by Sommerfeld. Of course, he wanted to improve it. In 1924 Heisenberg became assistant to Max Born (1882 - 1970) in Göttingen. The breakthrough came a short time, in 1925 on the island of Helgoland, where he cured his hay fever. He explained the frequencies of the spectral lines, including their intensities, using so-called matrices. He published his theory in 1925 and his boss Max Born and Pascual Jordan (1902 - 1980). This is considered to be the first quantum theory and is called matrix mechanics. I will not explain it in more detail because it's not very clear. And because there is an alternative mathematically equivalent to it.

It enjoys much greater acceptance because it is easier to handle. It is called wave mechanics and was developed in 1926, just one year after matrix mechanics, by the Austrian physicist Erwin Schrödinger (1887 - 1961).

THE SCHRÖDINGER EQUATION

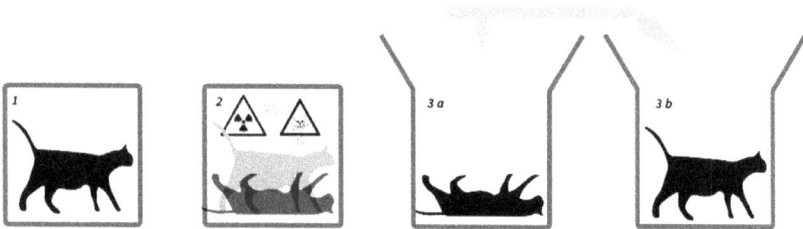

Before we come to the wave mechanics of Erwin Schrödinger, we must talk about the French physicist Louis de Broglie (1892 - 1987).

In his doctoral thesis, which he completed in 1924, he made a bold proposal. As explained in the penultimate area, wave-particle dualism was a characteristic exclusively of electromagnetic radiation.

Why, so de Broglie, shouldn't it also apply to matter? So why should matter not also have a wave character in addition to its real particle character? The examination board at the famous Sorbonne University in Paris was unsure whether it could approve it and asked Einstein. He was deeply impressed so that de Broglie got his doctorate. However, he could not present any elaborated theory for the matter waves.

Erwin Schrödinger then succeeded. In 1926 he introduced the equation named after him.

The circumstances surrounding its discovery are unusual. It is said that Schrödinger has discovered it at the end of 1925 in Arosa, where he was with his lover.

The Schrödinger equation is at the center of wave mechanics. As already stated, it is mathematically equivalent to Heisenberg's matrix mechanics. But it is preferred because it is much more user-friendly.

There is a third version, more abstract, developed by the English physicist Paul Dirac.

All three versions together form the non-relativistic quantum theory called quantum mechanics. As you rightly suspect, there is also a relativistic version.

The Schrödinger equation is not an ordinary wave equation, as it is used, for example, to describe water or sound waves. But mathematically, it is very similar to a "real" wave equation. Schrödinger could not explain why it is not identical. He had developed it more out of intuition. According to the motto: What could a wave equation for electrons look like? This can also be called creativity. Very often, in the history of quantum theory, there was no rigorous derivation. It was more of a trial and error until the equations that produced the desired result were found. Strangely, a theory of such precision could emerge from this.

However, as I will explain in detail, the theory is also mastered by problems that have not yet been solved. The solutions of the Schrödinger equation are the so-called wave functions. It was only with them that the stability of the atoms could be convincingly explained. Let us consider the simplest atom, the hydrogen atom. It consists of a proton as the nucleus and an electron moving around it.

EINSTEIN'S RELATIVITY

In 1907, only two years after developing the theory of special relativity, Einstein had the idea that he would describe as "the happiest of his entire life." In this inner vision, what would be revealed as the essential physical basis of general relativity appeared to him, even if it would take him almost ten years to elaborate the theory
mathematically?
Einstein realized that "if a man falls freely, he would not feel his weight." Even the expression "free fall" is telling: though one is always attached
to a gravitational field, attracted to the Earth from the perspective of Newtonian theory, one finds freedom when one is falling.
It is this freedom that those who pursue free-falling as a hobby seek to find and to feel, even if it is only partly due to air resistance.
It is, of course, astronauts in "weightlessness" who truly experience over a long period this feeling of no longer having any weight, of no longer being subject to the force of the Earth's attraction.
Nevertheless, the great idea of Einstein was the understanding that, if we jump up, during the brief moment of our jump, we experience this "weightlessness." In more words, there is no difference in principle between a vessel in orbit around the Earth and a ball which we throw here on Earth: both are in free fall; both are, for the duration of their motion, satellites of the Earth.

THE EQUIVALENCE PRINCIPLE

Understanding this universal phenomenon led Einstein to formulate the equivalence principle, according to which a gravitational field is locally equivalent to a field of acceleration. To obtain this principle, he drew upon a fundamental property of gravitational fields already brought to light by Galileo and included in Newton's equations: the acceleration communicated to a body by a gravitational field is independent of its mass.

After the development of special relativity, the need to generalize the theory seemed inevitable for multiple reasons. Relativist unification was far from complete. If the mechanics of free particles and electrodynamics finally satisfied the same laws, it was not the case for Newton's theory of universal gravitation, otherwise the top showpiece of classical physics. The equations of Newton are invariant under the classical transformation of Galileo, but not under those of Lorentz. Thus physics remained split in two, in contradiction with the principle of relativity, which necessitates the validity of the same fundamental laws in all situations.

Moreover, the Newtonian theory is based on certain presuppositions in contradiction with the principle of relativity: it is so with the concept of Newtonian force, which acts at a distance by propagating instantaneously at an infinite speed. The construction of a relativist theory of gravitation thus seemed to Einstein (and other physicists) a logical necessity.

Another problem was just as serious: the relativist approach explicitly gives itself the problem of changes in reference systems and their influence on the form of physical laws. But the answer provided by special relativity is only partial.

It only considers frames of reference in uniform translation, at constant speeds concerning one another. However, the real world constantly shows us rotations and accelerations, from the fact of the multiple forces which are at work (such as gravity), or inversely, causing new forces (such as the forces of inertia).

What are the laws of transformation in the case of accelerated frames of reference? Why would such frames of reference not be as valid for writing the laws of physics as inertial frames of reference? The answer is that such a question requires a generalization of special relativity.

The originality of Einstein's approach had been, in particular, to bring together two problems, that of constructing a relativist theory of gravitation and that of generalizing relativity to non-inertial systems, into a single endeavor. The equivalence principle made this unity of approach possible: if the field of acceleration and gravitational field are locally indistinguishable, the two problems of describing changes in the coordinate systems, including those which are accelerated and those which are subject to a gravitational field, boil down to a single problem. But such an approach is not reducible to "making relativist" Newtonian gravitation. While certain physicists could hope, at the time, that the problem of Newton's theory could be solved by a simple reformulation, by introducing a force that propagated at the speed of light, it is the entire framework of classical physics that Einstein proposed to reconstruct with general relativity. Better yet, it was a new type of theory which he developed for the first time: a theory of a framework (curved space-time, now a dynamic variable) in connection with its contents, and no longer only a theory of "objects" in a rigid preexisting framework (as was Newton's absolute space).

Why such a radical choice? Doubtless, because special relativity itself was unsatisfactory on at least one essential point: the space-time which characterizes it, even if it includes in its description space and time which is no longer absolute taken individually, remains absolute when taken as a four-dimensional "object." However, inspired by the ideas of Ernst Mach, Einstein had come to think that an absolute space-time could have no physical meaning, but rather, that its geometry should be in correspondence with its material and energetic contents. Thus, a reflection on the problem of inertial forces, which had caused Newton to introduce absolute space, led Einstein to the opposite conclusion.

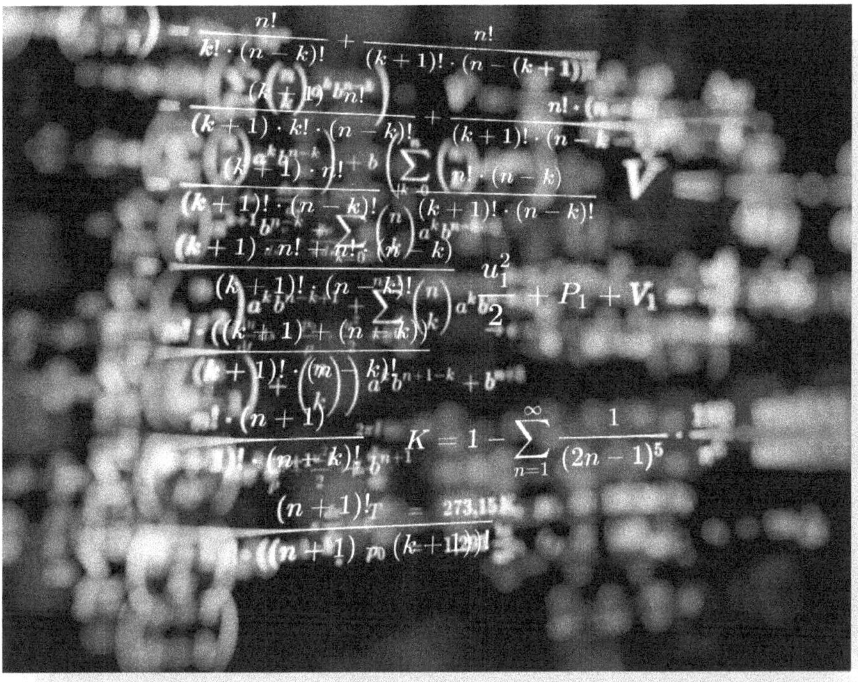

THE PROBLEM OF INERTIAL FORCES

The existence of inertial forces acutely poses the problem of the absolute or relative nature of motion and, ultimately, of space-time. The ideas of Mach in this area had a deep influence on Einstein. For Mach, the relativity of motion did not apply solely to uniform motion in translation; rather, all motion of whatever sort was by essence relative (Poincaré and, long before him, Huygens had arrived at the same conclusions).

This proposition can seem in contradiction with the facts. If it is clear, since Galileo, that it is impossible to characterize the state of the inertial motion of a body in an absolute manner (only the speed of a body concerning another has physical meaning), it seems different in the case of accelerated motions. Thus, when one considers a body turning about itself, the existence of its rotational motion seems to be able to be felt in a manner intrinsic to the body. No other body of reference is needed: it is enough to verify whether or not a centrifugal force appears, which tends to deform the rotating body.

In reconsidering the thought experiment of Galileo's ship, the difference between inertial movement and rotational motion becomes heightened. No experiment conducted in the cabin of a ship traveling in uniform and rectilinear motion concerning the Earth is capable of determining the existence of the boat's movement: as Galileo understood, "motion is like nothing." Relative motion can only be determined by opening a porthole in the cabin and watching the shore pass by. But now, if the boat accelerates or turns about itself, all the objects present in the cabin will be pushed toward the walls. The experimenter will know that there is movement without having to look outside. Thus, accelerated motion seems definable by a purely local experiment.

It is such an argument that caused Newton to allow that one can define an absolute space, in opposition with Leibniz (then Mach), for whom defining a space independently of the objects it contains could not have meaning.

Mach proposed a solution to the problem completely different from Newton's. Starting from the principle of relativity of all motion, he arrived at the natural conclusion that the turning body, within which there appear inertial forces, must turn not concerning a certain absolute space, but concerning other material bodies. Which ones? It cannot be tight bodies, of which the fluctuations of distribution would provoke observable fluctuations of inertial systems. This is unacceptable since it is easy to verify the coherence of these systems over great distances. Thus, if we look, motionless concerning the Earth, at the night sky, we do not see the stars turning.

Nevertheless, if we turn about ourselves, we feel our arms spreading out due to inertial forces, and, in raising our eyes toward the sky, we can see it turn. This was the initial observation of Mach: it is within the same frame of reference that the arms are raised, and the sky turns, and this will be true for two points of the Earth separated by thousands of kilometers. Mach suggested, then, that the common frame of reference is determined by the entirety of the foreign matter of bodies "at infinity," of which the cumulative gravitational influence would be at the origin of inertial forces. In other words, the body would turn concerning a frame of reference, not absolute, but universal. An absolute motion would be defined in itself, independently of all objects. However, Mach argued, all motion is relative, remaining defined for an "object," even if this object is the universe in its entirety.

The solution proposed by Einstein, that of the equivalence principle and general relativity, incorporates some of these ideas while ultimately distancing himself from the principle of Mach, even though his premises were identical. The supply of matter and energy in the whole of the universe determines the geometric structure of space-time. Then the movements of bodies are brought about within the framework of this geometry tied to matter.

RELATIVITY OF GRAVITY

Let us now return to Einstein's great idea in 1907. If an observer descends in free fall within a gravitational field, they no longer feel their weight, which means that they no longer feel the existence of this field itself.

This remark, which can now seem obvious to us—we have all seen, on television or in movies, astronauts in weightlessness floating in their ship, and the objects that they drop going away from them at a constant speed—is nevertheless revolutionary, for it implies that gravity does not exist in itself, that its very existence depends on the choice of a frame of reference.

He thus distanced himself from the former concept of gravity. What can be more absolute than a gravitational field in the Newtonian model? Gravity had been recognized by Newton as universal; here indeed was a physical phenomenon of which the existence does not seem to be able to depend on such and such a condition of observation.

However, if we allow an enclosed area to fall freely within a gravitational field, and then put in motion a body at a certain velocity to this area, the body will move in a straight line at a constant speed for the walls of the enclosure; a body initially immobile (again, to the walls) will stay thus during the movement of the enclosure's fall.

In other words, all experiments that we can perform there would confirm that we are in an inertial frame of reference! Thus, gravity, however universal it is, can be canceled out solely by a judicious choice of the coordinate system. What Einstein understood in 1907 was that even the existence of gravity was relative to the choice of the coordinate system.

THE DOUBLE-SLIT EXPERIMENT

Imagine an "electron gun" that fires electrons towards a wall with two holes (or slits) that are at equal distance (D) away from it, and equal distance (D') away from the center of the wall (Figure 1.0).
The electron gun is mounted on a turret, which moves back and forth from side to side, much like an oscillating fan. Given this motion, it's clear that we are not aiming the electrons at the holes; instead, they are simply being fired very much in a random fashion. The caves themselves are the same size and just big enough to let an electron through.

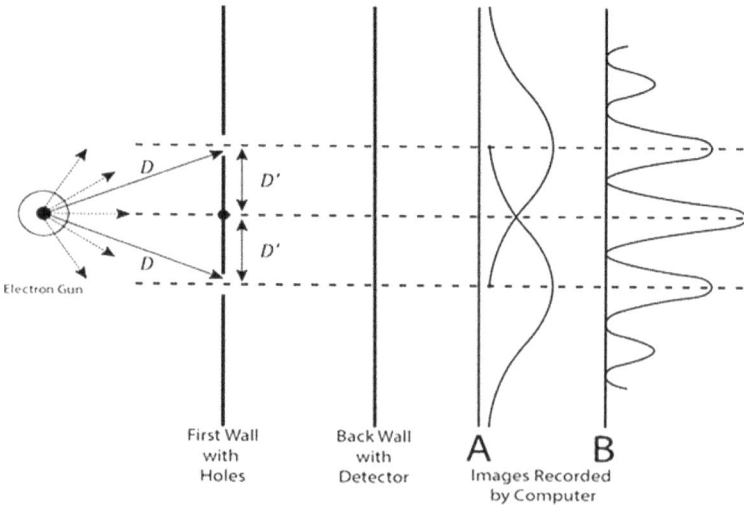

Figure 1.0. An "electron gun" fires at random upon a wall containing two holes that are at equal distance (D) away from it, and equal distance (D') away from the center of the wall. The electrons are stopped at

the back border, where a detector records their positions and sends this information to a computer. Image A is the distribution we get when we place sensors beside each hole to observe a passing electron. Here, we see no interference pattern, and in fact, we get the results we anticipated for an electron acting strictly as a particle, in which case it merely passes through one hole or the other. However, Image B is the distribution we get when there are no detectors present. Here, we do see an interference pattern as electrons pass through the holes.

As the electrons head towards the holes, some of them will pass through, and some of them won't. The passing electrons continue on their way until they end up hitting yet another wall located much farther down that acts as a backstop. At this back wall, the final position of each electron is recorded by a detector, which then sends this information to a computer for further processing.

As we keep terminating an ever-increasing number of electrons (we need to get great measurements), an ever-increasing number of electrons go through and hit the back divider. From the development of all the many-electron positions, the PC can make an example or dissemination. On the off chance that our measurements are adequate, at that point from this dissemination, we gain proficiency with the likelihood of finding an electron at a given situation on the back divider when terminated arbitrarily at the two openings. All in all, what does the dissemination resemble?

Before we acquire this, let's take a moment to try and anticipate the results. If an electron acts strictly as a particle, we would reasonably expect that it passes through one hole or the other. Moreover, an electron passing through a hole will either "bump" off of the side or edge, or will pass straight through unscathed. If it passes straight through, we'll find it directly behind the hole – at the "center," so to speak – when it hits the back wall, whereas if it gets bumped, we'll find it hits some distance farther away on either side of center. With all this in mind, we anticipate the distribution for a given hole to be such that the maximum number of hits occurs directly at the center, while farther away from there, the number of hits steadily decreases.

Lastly, the distribution will look the same on both sides of the center. In other words, it will be symmetric.

Ok, we have a pretty clear picture of what we'll see. But we experiment and find that the resulting distribution on the computer screen is nothing like what we imagined. Instead, we find a distribution with the maximum located between the two holes – it's not even located at the center of either hole! The distribution is still symmetric on each side of this maximum (so at least there's that), but we don't see the steady decrease in the number of hits that we had envisioned as we move away. Instead, on either side, we find peaks where the number of hits is high, and then from these peaks, there's a steady drop-off to zero, where not a single electron shows up. What happened?

Well, in our foresight, we assumed that an electron behaves as a particle, but we really should have known better since all quantum particles exhibit wave-particle duality. In short, the distribution formed by the collection of the many-electron positions is showing an interference pattern.

Earlier, we briefly talked about how interference can occur between waves. There must be waves associated with our electrons that are causing this interference pattern.

What are they? Recall that the quantum probability will determine the position of each electron at the back wall, as we mentioned earlier. In turn, the quantum probability is given by (the absolute square of the) wave function; it looks like we've found the "wave" causing the interference.

Let's try to recognize this in more detail. Instead of shooting many electrons all at once towards the holes, let's just shoot one electron at a time. Initially, we notice that shortly after firing off an electron, it arrives at the back wall, and its position is detected. So far, so good. However, as we continue to shoot individual electrons at the holes, we noticed something quite peculiar. Eventually, we end up with the same interference pattern that we saw before when we were firing many electrons. In other words, it does not matter if we fire several electrons at once or one at a time; the same interference pattern appears! This means that a single electron – as it encounters the two holes – ends up interfering with itself.

This seems so odd that we decide to perform one last experiment to get to the bottom of things. Beside each hole, we place a detector that will record an electron passing by it. Surely, this will shed some light on the strange results we are getting. Once again, we shoot one electron at a time towards the holes, over and over, until we can see the distribution on the computer screen.

This time we find that the interference pattern has totally disappeared, and instead, we are left with the distribution of electron positions that we had anticipated in the first place! In other words, when we're not looking at the holes (with our detectors), a single electron incurs interference. Still, when we do look, we find that the electron passes through either one hole or the other, and the interference pattern completely disappears.

These experiments illustrate the very essence of quantum mechanics. We see an electron acts as a particle when it hits the back wall and is detected by the detector as a localized entity. Still, somewhere in between, there's interference due to its wave nature and its "interaction" with both holes at the same time. This wave nature is intimately tied to the quantum probability of finding the electron at a particular position on the back wall, which ultimately leads to the distribution of hits we see. If we attempt to determine exactly where an electron will end up on the back wall by trying to see which of the holes it passes through, the whole thing falls apart, and the interference disappears altogether.

Although we chose to do our experiment with electrons, all quantum particles show this type of weird behavior. If all this seems like more science fiction than actual science to you, you're not alone. The physical consequences of quantum mechanics are, simply put, plain strange by comparison to our everyday experience.

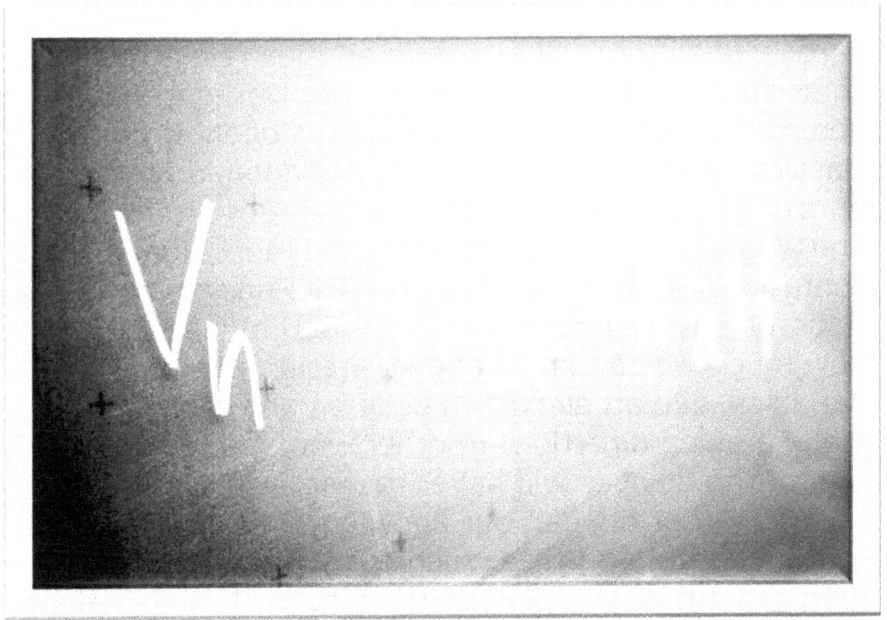

A Game of Chance

Quantum theory and its successor, quantum mechanics, rattled the very core of our understanding of the physical world we live in. Energy, light, atoms, and matter – all of these significant players came under heavy scrutiny. Even the very concepts that we use to describe some of these, like "wave" and "particle," were pushed to their breaking points, forcing us to now accept the existence of wave-particle duality for all quantum entities (electrons, light, and the like). And as if all that weren't enough, determinism itself, which had always been a part of classical physics, now had to be abandoned for uncertainty and an overarching quantum probability. These latter notions often cause the most significant confusion. Heisenberg's Uncertainty Principle defines a rigorous physical restriction (imposed by nature) on how much we can know about certain pairs of variables, like an electron's position and momentum along a given direction. In other words, having better measuring devices will never reconcile this inherent uncertainty or deepen our knowledge. It also means that a quantum particle, such as an electron, simply doesn't have a well-defined trajectory. It moves along. Rather, it "moves" between quantum states according to the quantum probability, which is related to Schrödinger's wave function. Bohr's atom with its "jumping electrons" – for all that it lacks – illustrates this quite well, much to Schrödinger's dismay. While the quantum probability is reminiscent of the classical probabilities of Maxwell and Boltzmann, nothing could be further from the truth.

For a while, the latter are self-imposed to ease the burden of complicated mathematics; they still preserve the underlying determinism so dear to classical physics. In contrast, the quantum probability is an unequivocal affirmation by nature against determinism altogether. Indeed, the reckless abandon of the well-oiled "world machine" for nothing short of a "game of chance" undoubtedly poses the greatest challenge to one's sensibilities. Nonetheless, by all current accounts, quantum mechanics, with all its "weirdness" and probabilistic overtones, has endured the test of time.

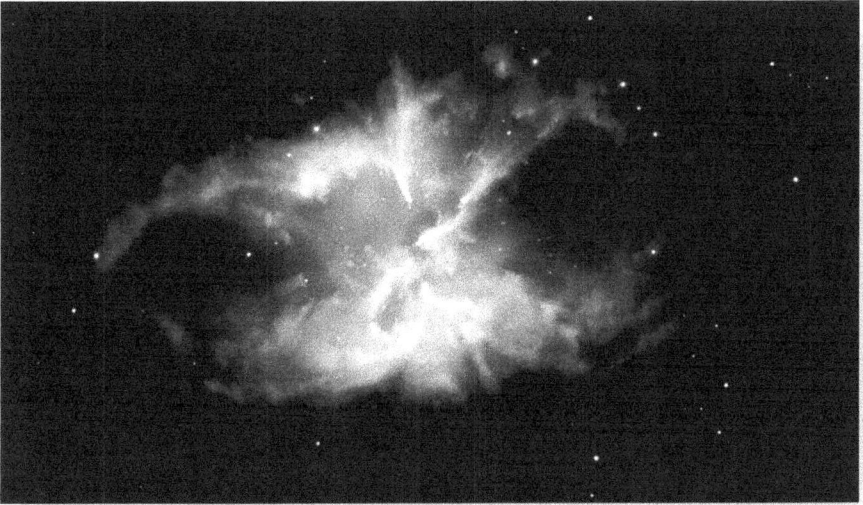

Black Body Emission

The wave hypothesis of light was the overall light hypothesis during the 1800s. This hypothesis was captured by Maxwell's conditions   and   bested Newton's   corpuscular theory. Nonetheless, the hypothesis was challenged by how it clarified warm radiation, which is electromagnetic radiation emitted by objects dependent on their temperature. So how might somebody test or distinguish friendly radiation? Likewise, with some other quantum testing, the capacity to make a fruitful trial relies upon whether there exist the identifiers and mediums accessible to both measure and lead the problem itself.

For instance, researchers can test for warm radiation by setting up a device to distinguish radiation from an item dependent on a particular temperature, indicated by T1. Living, breathing people give off radiation every which way, so to have the option to quantify it adequately, protecting must be utilized, so the radiation is analyzed as a limited pillar. By using the shielding, a scientist can create the conditions desired to focus a narrow beam. Therefore, a scientist can begin to create suitable surroundings for this experiment.

To make this thin bar, a researcher utilizes a dispersive medium, for example, a crystal, situated between the body or item, delivering the radiation and the radiation indicator. This encourages the radiation frequencies to be diffused at a point. At that point, the indicator decides a particular range or edge, basically the limited pillar.

This pillar has contemplated a portrayal of the all-out power of the electromagnetic radiation all through all the frequencies.

Yet, how does all the force interpret across frequencies, and how would we diminish the different qualities to make functional conditions?
So we should clarify a couple of central issues. One thing to comment is that the power per unit of a frequency stretch is intended to as radiancy. Math documentation underpins us to decrease different qualities to zero and make the succeeding condition: $dI = R(\lambda)\, d\lambda$. Utilizing the crystal, a researcher can recognize dI or the absolute force over all frequencies, so one can characterize radiancy for any frequency by working in reverse through the condition. Presently we should look at how we can manufacture an information base of sorts for frequency versus radiancy bends.

Remember that each information base is worked through an assortment of investigations. Comprehend that researchers ordinarily analyze, again and again, building up a heap of information that structures different extents. When working with these reaches, one can start to manufacture a superior comprehension of how much radiation will happen from a particular item, yet in addition to how extreme it will be at some random temperature.

For example, one can suspect that as the absolute power transmitted multiplications as we rise or fall the temperature. However, when we consider at the frequency with the most extreme radiancy, we find that the reverse happens; that is, with that particular frequency, the force will go down as the temperature increments. Consequently, as the temperature goes up, frequencies can change their radiation power, yet the general radiation force will keep on expanding with the weather.

So if the temperature is going down, at that point, the most incredible power of an individual frequency will go up. The by and large or absolute control of the article will go down, compared with the temperature. For this situation, temperature assumes a necessary

function in how the test results will play out. Expanding or diminishing the temperature can influence the outcomes. However, researchers likewise expect specific effects inevitably. So on the off chance that they increment the weather, however, get unexpected results in comparison to what they were expecting, the researcher will search for any possible defects in the trial arrangement itself. In any case, with each endeavor, they assemble an extraordinary storage facility of information about how these different parts connect.

Once more, we go consequently to how to ascertain generally when light reflects off endless things. How would you make the edge, making sure that you precisely measure your limited shaft?
A straightforward strategy to do this is to quit taking a gander at the light. Instead, center around the article that does not uncover it. Light rehashes off of articles; however, researchers will achieve this test recognizing a blackbody or an item that doesn't display whichever light by any stretch of the imagination.

Something else, the analysis goes into an endeavoring jumping of what is being tried.
Finishing this trial constrains a crate, if conceivable, metal, with a small opening. If or when light knocks the gap, it goes into the case; however, it won't hop back out. As an answer, the hole, not the crate, is the blackbody of the test. Any radiation distinguished outside of the opening is a radiation test of the measure of radiation in the case.

Researchers examine this proof to sympathize with what's happening inside the bundle.
The underlying thing to be remarkable is that the metal box is being expended to stop the electric field at each mass of the crate, starting a hub of electromagnetic vitality at every one of the dividers. Hence standing electromagnetic waves are controlled inside the case.

Second, the number of endless waves with their constant frequencies inside a characterized go, containing a condition that returns into balance the volume of the case. By investigating the standing waves and afterward following this condition, it very well may be venturing into three measurements. As we have referenced with measurements, numerous speculations take growing from three sizes into some more, yet they might not have reliable testing alternatives.

Third, traditional thermodynamics contributes an essential truth: the radiation in the case is in warm harmony with the dividers of the issue at a distinct warmth. The radiation inside the container is assimilated and re-transmitted from the walls, regularly delivering wavering inside the radiation's event.

The warm, unique vitality of these wavering particles is quiet consonant oscillators, so the typical kinematic dash coordinates the mean possible spirit. Subsequently, each wave gives the whole life of the radiation in the container.

Fourth, vitality thickness is related to brilliance. Vitality thickness is depicted as the vitality per unit volume inside the relationship. The estimation of this is top-notch by the measure of radiation going through a segment of the surface region with a depression.

Great material science, as encapsulated by the Rayleigh-Jeans equation, neglected to anticipate the genuine aftereffects of these analyses, basically because exemplary material science ignored to represent shorter frequencies. At more extended frequencies, the Rayleigh-Jeans equation all the more firmly concurred the recognized information. This disappointment was spoken to as the bright calamity. In the mid-1900, this was a huge issue since it raised doubt about such unique ideas as thermodynamics and electromagnetics as an element of that equality.

Verifiably, this is the place quantum material science is inferred into play. Nearly, Max Planck devoured quanta to create what might be set up as a free lot of solidarity. Consequently, the quanta would be proportionate to the vitality recurrence. With this hypothesis, no vertical wave could have more vitality than kT, and afterward, a high radiation event would be topped, deciphering the bright demolition. At long last, recurrence assigns the dynamism of every quantum, were a similar standard.

—

While this stemmed in a condition that appropriates the information of the analyses properly, however, it wasn't as shocking as the Rayleigh- Jeans detailing. This recipe built up the beginning stage of quantum material science as we probably are aware of it today. Einstein even confined it as a focal standard of the electromagnetic region, while Planck had, in the past, utilized it just to answer the subject of a solitary analysis. While it acquired researchers some time to get used to what exactly is presently known as Planck's Constant, it is currently viewed as a basic aspect of the quantum material science or quantum mechanics.

This was only one aspect of the huge cluster of analyses that characterize quantum material science. Another early examination is working together with wave-molecule duality, a test that was known as the photoelectric impact.

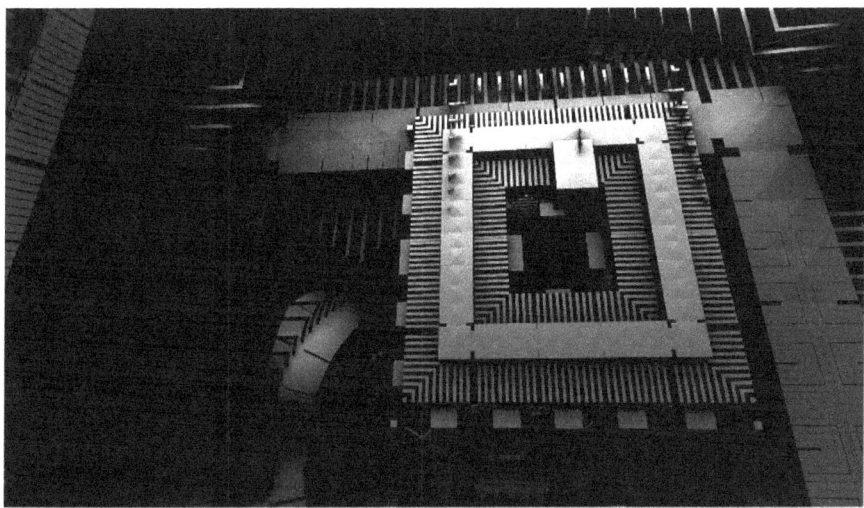

Quantum Reality

An isolated quantum system is always in a superposition state, a state of potentiality before it is observed. It has no inherent characteristics except charge and mass. Position and velocity do not apply to a quantum system before measurement because such properties are produced as a result of the measurement.

Isolated quantum objects, such as an electron, behave like ideas: they exist in an abstract mathematical space. What is observed of these abstract objects during the measurement process is their expressions/projections in our measurement apparatus; it is the apparatus that plays the role of a basis in quantum physics.

The expression we get on the apparatus is the representation of the quantum object in terms of the capabilities of the measurement apparatus. If the apparatus measures the position of quantum objects, then by using this apparatus on an electron, we obtain the representation of our electron in the position space. Thus, the position we assign to an electron is only its expression in our apparatus and not an objective property of the electron itself—in the same way, that the letters of the word apple do not belong to the idea "apple."

As it is meaningless to assign letters to an idea, it is equally meaningless to assign a position or velocity to an electron in itself. This is the point of departure for quantum physics. According to quantum theory, what has been thought of as objective reality has no real, independent existence in itself; it is a mere expression in the observer who plays the role of a representative basis.

Philosophers divided the qualities of an object into primary qualities and secondary qualities.

Primary qualities of an object are defined to be those that are independent of any observer; whether the object is lost in deep space or it is being observed, these primary qualities are not affected because they exist in the object itself. Examples are shape, size, and according to classical physics, also the state of motion of the object, which includes position and velocity.

Secondary qualities of an object are defined to be those that depend on the observer; these qualities exist whenever the object is being observed or experienced in some fashion. In other words, secondary qualities exist in our experience of the object and not in the object itself. Examples are color, taste, smell, apparent size and shape, etc. Color, for instance, does not exist in the object; the object is made up of atoms and electrons that are colorless and tasteless; what we experience as color comes from our brain's interpretation of the frequencies of the light emitted by the atoms of the object.

Although secondary qualities do not exist in the object itself, the object has to be present for us to experience its secondary qualities. These qualities, therefore, are neither in the object nor in the subject; instead, they belong to and exist in the interface of interaction. They only appear during the event in which the subject interacts with the object.

What happened in the advent of quantum physics was that the primary, objective qualities of our world turned out to be only secondary qualities. Based on experimental findings, quantum physics showed that qualities such as position, velocity, energy, and hence the state of motion of an object do not exist in the object itself; they belong in the interface of interaction, which is now the measurement apparatus, namely, the representative basis. Quantum physics undermined the idea of Realism.

This paradigm shift undermining the objectivity of our world is what makes quantum physics counterintuitive. Once we accept that an

isolated quantum object possesses no specific location or velocity or energy, then we cannot even imagine such an isolated system—not so much because we do not know what it looks like, but rather because a quantum object is not a kind of object that looks like at all, let alone looking like something. The quantum object has no look, and hence, it is not something to look at!

The properties of quantum objects, and those of the objective reality that is essentially quantum mechanical, exist only during observation and experience. The world exists only in our experience. Belief in the independent existence of the world is itself something inside experience.

The world does not possess any objective property because the world is not a kind of thing to possess anything whatsoever. The world is just an appearance. Quantum particles—such as electrons, protons, and all other subatomic particles—are only mathematical constructs, like Plato's regular solids. A measurement apparatus displays only some digits on it, which are to be interpreted by humans as something meaningful. The world of phenomena, the world we know, is not something of which we have experienced; it is not something that exists apart from experience; it exists only within observation and experience.

What we perceive is not the world as an entity, but what appears on the interface of interaction. Werner Heisenberg writes in his Physics and Philosophy: More than being just a description of the subatomic world, quantum physics is a statement about the classical picture of the world. It exposes the inconsistencies and paradoxes inherent in our everyday view of the world as something material, deterministic, and indifferent to consciousness. For instance, Heisenberg's Uncertainty Principle rejects the possibility of material solidity.
Now let us see how the superstitious ideas of solidity and material constitution originated in the first place and why they are false.

The classical view of the world is rooted in our conscious experience. We experience objects as solid, liquid, or gaseous; thus, we have these main distinctions in classical physics.
Objects of our experience appear to enjoy an independent existence subject to certain causal relationships; hence, the objective and deterministic aspects of the classical view. But if we pay thorough attention to the same experiences within which these ideas arose in the first place, we realize the following: All objects, and the whole of our world, are first given to us in experience, but the experience is a temporal stream of consciousness; the solid appearance of objects is itself something experienced in time. Solidity is itself a phenomenon that belongs in the temporal flow of experience. (If I don't use the phrase my experience, it is because the "I" is itself something experienced; it belongs in the content of experience and not outside it. Experience is not something I have because I am myself something experienced.)
The world and its objects are phenomena constituted within the temporal flow of subjective experience.

—

It is in time that they are what they are. Objects are first and foremost temporal entities, for it is in and through time, as flow, that their existence has any sense. In other words, objects are first extended in time before they are extended in space. It is within the primordial river of time that reality makes its first appearance and acquires its objectivity.

It is essential to distinguish between something that flows and something that exists within the flow. The flow of a river is the flow of water molecules; these molecules exist even when they are not flowing like the river; flow is only one possible state of motion for them.

However, the temporal flow of experience within which objective reality is constituted is essentially different from the flow in the sense of a river. Temporal flow is a state of constant flux within which everything is constituted; nothing exists outside this flux because existence is itself something constituted in time.

The fact that reality is essentially a temporal phenomenon is not calculated into our classical view of reality. The Uncertainty Principle in quantum theory is a technical restatement of the nonphysical but temporal constitution of phenomena. This is why a point-like particle makes no sense in this primordial flow that we call the Heraclitean Flux.

An analogy can help us see how something can exist only within a flow: imagine that you see something that appears to be a circle from afar; when you get close to it, you see a bright circular object hanging in space. Although there appears to exist a circular object, the bright circle is nothing but the trace of a ball of fire that is being rotated very fast by an invisible fire dancer.

The circular object appearing to exist as something in space is a mere appearance existing only within the fire flow created by the dancer. If the rotation stops, the circle no more appears. Notice that the circle never existed; to begin with, it was just a ball of fire only appearing to be a circle due to some constant flux.

Now, if you hit such a circle with a stick, it will lose its shape and disappear altogether, because you disturbed the very flow within which it existed as appearance. This is very similar to what happens in the quantum world when the measurement is performed. The act of measurement destroys the very appearance under investigation.

We see that some of the strange features of quantum physics are not so strange if we correct our classical view of the world by deeply reflecting on experiences within which we have the world in the first place. If objective knowledge has to begin with experience, end with experience, and never lose its rootedness in experience, then our slightest ignorance about the vital role of experience in the constitution of reality leads to a knowledge detached from reality.

The facts of the microscopic world articulated in quantum physics are experimental references to the underlying reality as a Heraclitean Flux, which is essentially a temporal-subjective reality rather than a spatial- objective speculative fiction.

Mathematics the Language of Physics

The Role Of Mathematics In Physics

First, Two Comments:

1.   Roland Omnès, an esteemed Professor of Theoretical Physics in France and graduate of the élite Ecole Normale Supérieur in Paris, states at the beginning of his 2002 "Quantum Philosophy":
If I had to name the greatest thinker of all times, I would say, without hesitation, Pythagoras, who lived on the Greek island of Samos, 6th Century B.C.) ....
He said that numbers rule the world.

2.   To the aspiring Scientist or Mathematician:
Follow the numbers, no matter where they lead you. They hold the truth to the Universe.

The important working relationship between the world of physics and the world of mathematics is both synergistic (work together or enhance one another) and symbiotic (two different systems live together). This relationship is a key operating feature in the advancement of science.

First, note that there are essentially two branches of mathematics: Theoretical Mathematics and Applied Mathematics. The discussion will be about Applied

Mathematics. But, in deference and respect to the Theoretical Mathematician, I would first like to explain what a theoretical mathematician does.

Examples of Where Applied Mathematics Relates To The World

Many people in the world are trying to understand what happens in the physical world. For example, there are:

•   Weather and climate analysts, who seek to understand what makes and drives the various weather and climates.

•   Oceanographers who seek to understand currents and tides in the water.

•   Engineers who build buildings and calculate what forces of weather can or will knock down a structure.

•   Aeronautical engineers are determining the strength of materials.

•   NASA engineers designing space-flights trajectory to the planet Mars.

•   An automotive engineer designing a new type of contour (curvature) for the hood of a new sports car.

•   Physicists are analyzing a substance when the temperature of the substance approaches zero degrees Kelvin or analyzing how electricity flows within the substance, also near zero degrees Kelvin.

- Computer architects are trying to reduce the number of components (molecular or smaller) on a memory chip or a logic board.

- Physicists are trying to understand the inner features of Black Holes.
- Physicists are developing String Theory.

- Physicists are attempting to incorporate Quantum Gravity into the Standard Model.

- Physicists are trying to harness and apply Quantum Physics to a wide range of objects, including computers, communication systems, medical detection, monitoring systems.

- And thousands more.

These are all people who rely on mathematics to describe the process they are analyzing. Thus, that is why this sector of mathematics is called Applied Mathematics.

They first try to understand the process. When they achieve that, they seek to describe it by mathematical formulae, called equations. Sometimes the formulae are known and available; sometimes, there is no formula available. In the latter case, they try to derive it themselves— or they wait for a separate group of mathematicians to develop the formula. Frequently, the development of the mathematic equation provides the developer with additional insights into the physical process they are investigating. Mathematics is usually a stand-alone profession performed by mathematicians; they produce mathematical formulae. Since not many people 'buy' mathematical formula, mathematicians work in, or with, other professions such as being instructors and professors in colleges and universities, and in government and not-for- profit organizations (where mathematical studies are performed), and in industrial and research organizations.

A modern application is the development of a spectacular range of mathematical formulae to describe—and predict, the events and activities of the stock market, the bond market, the futures market, the real estate market, the credit market, the mutual fund market, and so on.

Many of us learned about Pythagoras, a Greek mathematician (circa 500 B.C.), who looked at a right triangle and applied an equation to it that we call the Pythagorean theorem. The theorem is that the sum of the square of the sides is equal to the square of the hypotenuse. That equation enables people to calculate the sides of a right triangle for all different size right triangles.

In the case of Pythagoras—or a fellow member, he would see a physical object and realized a mathematical formula could describe the object. (Just as Kepler did, but Kepler did it for the planets.)

There are many different, well-known equations, and they frequently are referred to by the name of the founder. For example, there is the Bessel Function, named after Frederick Bessel (1784-1846). Fortuitously, The Bessel Function can be used—and is used to describe the FM radio wave that we listen to on the FM band of our car, iPods, and home radios.

And then there is the example that always intrigued me. It has to do with a favorite mathematician of mine, Leonhard Euler (1707-1783), considered the greatest mathematician of the 18th Century. Engineers in the 19th and 20th Centuries were building tall buildings. They would put a steel beam vertically into the ground. It was held in the ground by concrete or some other type of 'building glue.' They also attached steel beams to other beams, extending their height. Or they would place a steel beam horizontally, supported by other beams.

Leonhard Euler had developed such a formula in about 1755. This was long before steel was even thought about. Thus, Euler had no idea what a steel beam was. But his formula was used one hundred fifty years to predict the bending of the beam if or when a force pushed acted on the beam. And civil engineers used the Euler equation as they created new buildings and new cities.

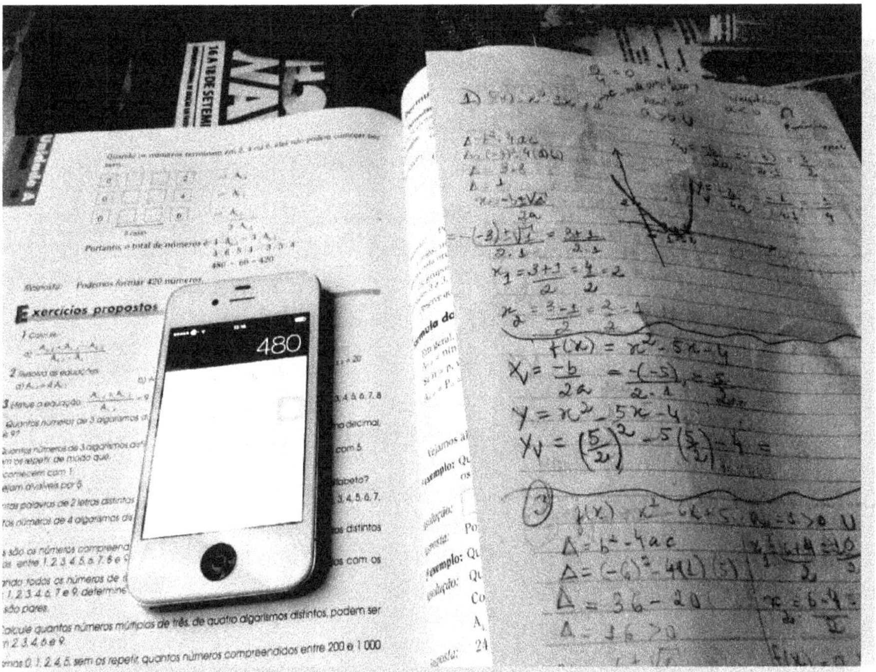
Exercícios propostos
480

THE EXAMPLE OF THE APPLICATION OF BERNHARD RIEMANN'S MATH:

Bernhard Riemann developed another set of 'exhilarating' equations in 1854.

These equations describe a geometry that was different from Euclid's two-dimensional geometry, which you and I had been taught in school. Riemann's geometry represented 'unflat,' undulating surfaces.

In the early 1910s, Albert Einstein had developed a theory of space and gravity. But he did not know how to express it mathematically. It was only after he learned (from his mathematician-friend, Marcel Grossman)about Riemann's geometry that he could describe his thoughts mathematically.

This application of Riemann's geometry became known as a crucial part of Einstein's Theory of General Relativity. Without that mathematical description, Einstein—or anyone else, could not have made calculations about gravity, space, and Universe for various situations.

I end this introduction with a quote from Albert Einstein: "The approach to a more profound knowledge of the basic principles of physics is tied up with the most intricate mathematical methods."

Three Outstanding Mathematicians: Leonhard Euler, Srinivasa Ramanujan, And John Von Neumann

I have long been aware of the extraordinary and interesting lives that many mathematicians have led—and would like to devote many more pages to their personal history and contributions to pure math. It relates to the world of physics. Since that is not realistic for a reading devoted to physics, I limit my discussion to only three vastly different mathematicians, Leonhard Euler, Srinivasa Ramanujan, and John von Neumann. The first is presented for his contributions to many areas of mathematics and physics, the second for his overall creativity and genius, and the third for his creativity and leadership in three quite different, modern arenas.

Leonhard Euler, The Most Prolific Mathematician of All And 'Master Violinist Of The Strings.'
The first time I learned about Leonhard Euler, the Swiss mathematician (1707-1783), was when I was taking a civil engineering course, examining the strength of steel beams. The professor described an equation that he attributed to Euler.
I list some of his varied contributions below and give a number to each of Euler's 'fields of endeavor.' I believe, however, that there are many more that are not mentioned here.

1.   Euler was responsible for "Combinatorial Analysis." An example of combinatorial analysis: How many ways can eight different items be taken from a group of 21 different items? Combinatorial Analysis is

fundamental to the development of statistical theories used in Quality Control, a key ingredient of the modern industrial age.

2.   Euler is considered the father of "Modern Graph Theory." Euler's graphs were what are called linear graphs, which he developed to solve problems. Today we all know of the 'salesperson problem,' which these linear graphs solve the problem of where a salesperson must visit a certain number of clients, and the routes are analyzed to minimize the trip—and thus, implicitly, to increase sales and profits.

3.   A branch of the mathematical discipline known as Calculus is "Differential Equations." It appears that there were two important phases in the development of Differential Equations. Differential Equations are the probably most widely used system of equations to express physical situations in which the parameters, time and changing time, appear, such as in liquids and gas flows. In a similar time-varying case, Schrödinger used differential equations to express his time-varying quantum equations.

4.   In the sector of physics known as Fluid Mechanics, Euler explicitly set forth the concept of the internal pressure in a liquid. He also developed the equations for the formulation of a three-dimensional description used for fluid flow. The study of the pressure and flow of liquids is important today in many of our common industrial systems. Examples include petroleum, natural gas, and water pipeline networks that crisscross our continent; similar pipeline networks interconnect European and Asian countries, enabling them access products across national borders; and there are similar pipeline networks within many of the world's cities, themselves.

A Creative Mathematician, Srinivasa Ramanujan—And A Genius (1887 – 1920)

Srinivasa Ramanujan was a mathematician who had a non-standard personal history.

He developed many theorems by himself, living in India, and many of these theorems were unknown in Europe and America. Some of his work may help us understand some of the basic problems and unsolved theories that challenge physicists today and in the future.

For example, in the part of physics called 'String Theory' people believe that Ramanujan's formula called the Elliptic Modular Function will help describe the theory further, mathematically.

John Von Neumann (1903 - 1957)

John Von Neumann (1903-1957), a Hungarian mathematician, distinguished himself from his peers— even in childhood, for having superior mental faculties.

For example, he had a photographic memory. He was able to memorize and recite back a page out of a phone pad in a few minutes.

Also, at the age of six, he could divide eight-digit numbers in his head. He was notable for having been involved in many aspects of modern physics and engineering.

Von Neumann published his first mathematical paper at the age of eighteen, in collaboration with his tutor, and decided to study mathematics in college. In 1926 he received his Ph.D. in mathematics with minors in physics and chemistry from the Pázmány Péter University in Budapest, Hungary.

What Are Your Quantum Thoughts?

Quantum understanding of the quantum thoughts is a set of Hypotheses suggesting that classical mechanics cannot explain consciousness. It may explain consciousness and posits that phenomena, such as superposition and entanglement, may play a significant role in the functioning of the brain. Assertions that understanding can seep a movement which assigns traits to quantum phenomena like no locality and the audience effect, with quantum mysticism.

History

Eugene Wigner developed the idea that quantum mechanics has something related to his mind's workings. He suggested the wave function collapses because of its interaction with all consciousness. Freeman Dyson contended that "thoughts, as exemplified by the ability to make decisions, are to some extent inherent in each electron." Philosophers and other physicists believed these arguments were unconvincing. Victor Stinger recognized quantum understanding as a "fantasy" with "no scientific basis" which "should take its place together with gods, unicorns and dragons" Quantum understanding is argued against by David Chalmers. He discusses quantum mechanics can relate to consciousness. Chalmers is doubtful that any physics could solve the problem of consciousness.

Quantum Mind Tactics

David Bohm

David Bohm saw relativity and theory as Contradictory, which indicated a level in the world. He maintained both quantum theory and pointed out that this level was suggested to represent order and an undivided wholeness as we encounter it, where originates the order of this world.

The proposed implicate order of Bohm applies both to consciousness and matter. He suggested it may clarify the connection between them. He saw matter and mind in the implicate order to our order as projections.

Bohm is maintained when the thing is looked at by us, we find. Bohm mentioned the experience of listening to songs. He considered the feeling of changes and motions which constitute our experience of audio derive from holding the current from the mind and the past.

PENROSE AND HAMEROFF

Theoretical physicist anesthesiologist and Roger Penrose Stuart Hamer collaborated to create the concept called Orchestrated Objective Reduction (Orch-OR). Hameroff and Penrose developed their thoughts and collaborated to generate Orch-OR from the early 1990s. They updated their vision and rechecked. Penrose's argument stemmed from theorems. In his very first reading on understanding, The Emperor's New Mind (1989), he contended that although a formal system can't prove its consistency, Gödel's improvable outcomes are provable by individual mathematicians. Penrose took this to imply that mathematicians aren't operating traditional proof systems, but rather, they are operating an algorithm. According to Xiao and Bringsjord, this line of reasoning relies on equivocation.

At precisely the same publication, Penrose wrote, "One may speculate, but that somewhere deep within the mind, cells should be discovered of single sensory sensitivity. If it turns out to be the situation, then quantum mechanics will be involved in brain activity." Penrose decided wave function collapse was that the Potential sole basis for a non- computable procedure. He suggested a sort of wave function collapse, which predicted its reduction and happened in isolation. He indicated when these become divided, they become unstable, and fall and every quantum superposition has its parcel of curvature. Penrose suggested that reduction signifies no randomness or algorithmic processing but rather an effect in geometry by comprehension derived and by the expansion that was.

Hameroff supplied a theory that microtubules will be hosts for quantum behavior. Microtubules are composed of protein dimer subunits.

The dimers might include pi electrons, and each has pockets. Tubulins have other areas that include pi indole rings. Hameroff suggested these electrons are enough to become entangled eventually.
He suggested the electrons could create a Bose.

Penrose Continues...

A good deal of what the mind does you can do on a pc. I am not saying that the whole mind's activity is different from everything you can do on a pc. I'm claiming that conscious activity is something. I am not saying that understanding is beyond mathematics, either--though I am saying that it is past the physics we understand today... My claim is that there needs to be something in physics that is a personality, and which we do not know yet, which is quite significant. It is not specific to our brains; it is out there in the universe. However, it plays with a function. It might need to maintain the bridge between classical and quantum levels of behavior --where quantum dimension comes in. W. Daniel Hillis reacted," Penrose has committed the classical error of placing people in the middle of their world. His argument is he cannot envision the mind might be as complicated as it is without getting some elixir brought from a principle of mathematics, so it has to involve that.

DAVID PEARCE

English savant David Pearce guards what he calls
genuineness vision ("the non-realist rawness attest
that the truth is experiential and the natural universe
is thoroughly portrayed by the conditions of science
just as their answers"), additionally contains guessed
that unitary cognizant personalities are physical
conditions of quantum intelligibility (neuronal
superposition's).
This guess is further, as indicated by Pearce,
manageable to misrepresentation, in contrast to
numerous speculations of understanding.
Additionally, Pearce has summed up an exploratory
equation portraying how the theory could be tried with
an issue wave interferometer to find neoclassical
obstruction examples of adrenal superposition toward
the beginning of warm decoherence. Pearce concedes
that his considerations are "exceptionally theoretical,
"unreasonable," and "extraordinary."

CRITICISM

Since Penrose and Pearce acknowledge in their discussions all these hypotheses of their quantum mind remain speculations, the ideas are not based on evidence until they make a forecast that's examined by experimentation.

According to Krauss, "It is a fact that quantum mechanics is very odd, and on extremely tiny scales for quick times, all kinds of bizarre things happen. And we could make quantum phenomena that are bizarre occur.

However, what quantum mechanics does not change about the world is, even if you would like to modify things, you still need to do something. You cannot change the world by considering it." The practice of analyzing the hypotheses with experiments is fraught with conceptual/theoretical, functional, and ethical issues.

Conceptual Issues

The concept that a quantum impact is essential for consciousness to work remains in the world of doctrine. Penrose suggests it is crucial, but other notions of consciousness don't signify it is required. By way of instance, Daniel Dennett suggested a concept called multiple drafts version that does not imply that quantum effects are necessary within his 1991 Consciousness Explained.

The philosophical debate on each side is not scientific evidence. However, a philosophical investigation can indicate crucial differences in the sorts of models and reveal which some kind of observed differences may be observed. But because there is no consensus among philosophers, supporting a quantum mind concept is necessary. You will find computers that are mainly designed to calculate using quantum mechanical results. Quantum computing is computing with quantum-mechanical phenomena, such as superposition and entanglement.

They are not the same as binary digital electronic computers based on transistors. Among the most significant challenges is removing or controlling quantum decoherence. This means typically isolating the system from its surroundings since interactions with the outside world cause the system to decoherer.

Quantum entanglement is a bodily phenomenon frequently invoked for quantum brain versions. This effect happens when groups or pairs of particles interact, so the quantum state of every particle cannot be described independently of another (s), even if a considerable space separates the particles.

Instead, a quantum state needs to be clarified for the entire system.

Measurements of physical attributes like position, momentum, spin, and polarization, conducted on entangled particles are seen to be connected.

PRACTICAL ISSUES

The demo of a human brain impact by experimentation is essential. Is there a way? Could a complex electronic personal computer be shown to be incapable of awareness? A quantum computer will demonstrate that quantum effects are wanted. Whatever the instance, quantum

computers or electronic computers could be constructed. These can show which sort of computer is capable of conscious thought. However, they don't exist, and no evaluation was demonstrated. Quantum mechanics is a model that may offer some accurate forecasts. Richard Feynman anticipated quantum electrodynamics, contingent upon the quantum mechanics formalism, "the gem of material science" for its very exact expectations of amounts like the strange attractive snapshot of the electron and the Lamb move of the vitality degrees of hydrogen.: Hence, the model may give a precise estimate about understanding that could confirm a quantum impact is incorporated. The proof is to seek out an experiment, which proves whether the brain is dependent upon quantum effects. It must demonstrate a difference between a computation leading to one, which entails quantum effects and a mind. The theoretical argument against the brain theory is that the assertion until they attained a scale where they might be helpful for processing, that quantum states from the mind would eliminate. Tag Mark elaborated on this supposition. His calculations suggest that quantum systems from the mind decoherer in timescales. No reaction by a mind has revealed reactions or benefits. Responses are on the order of milliseconds more than timescales.

The Penrose States

The issue with attempting to use quantum mechanics in the activity of the mind is that if it had been an issue of quantum neural signs, these neural signs would disturb the remainder of the substance from the reason to the extent that the quantum coherence would have lost very fast. Within an environment that cluttered and you could try to construct a quantum computer. Nerve signs need to be treated.

But if you move down to the degree of the microtubules, there is a very good possibility that you could get action. For my image, I want this action at the microtubules; the story needs to be a scale item that goes across large regions of the brain, from one microtubule to another but from one nerve cell to another. We want some type of action of a quantum character that is coupled that Hameroff asserts is happening across the microtubules.

There are paths of attack. One is on quantum theory on physics; also, there are strategies to get a modification of quantum mechanics and experiments which folks have begun to execute.

Ethical Issues

Will mindfulness, or even perception of yourself at the general condition, be cultivated by a traditional equal chip, or are quantum impacts essential to have a sentiment of "unity"? Following Lawrence Krauss, "You should be mindful if you hear something like, 'Quantum mechanics go along with you with the world'... or then again 'quantum mechanics connecting you along with all the fixings ' It is conceivable to begin to be incredulous that the speaker is attempting to utilize quantum mechanics to state generally which you could change the world by thinking about it." An abstract inclination isn't sufficient to create this assurance. Individuals don't incline how they do numerous capacities.

Because of Daniel Dennett, "Regarding this matter, everybody's an expert... yet, they accept they have a particular individual authority concerning the quintessence of their cognizant encounters which may disregard any hypothesis they find unsatisfactory." Doing trials to show quantum impacts requests experimentation on a human psyche since individuals are the main animals that may convey their experience.

QUANTUM DIMENSION

Quantum mechanics had its genesis in a simple experiment originated by Thomas Young, a British researcher of many fields of science and humanities, over two hundred years ago. The experiment requires only a light source, a board with two slits, and a screen on the other side to catch the light that passes through the slits:

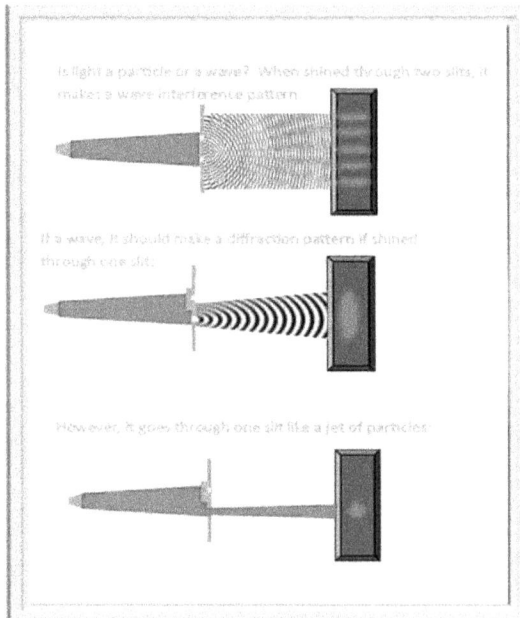

When Thomas Young reported the first "double-slit" experiment in 1801, the scientist who became the Lord High Chancellor of Great Britain decried it as "destitute of every species of merit" and "the unmanly and unfruitful pleasure of a boyish and prurient imagination."

What did Mr. Young do to provoke such outrage from a country noted for its culture of understatement? He showed that light manifested a dual particle/wave nature. But it wasn't the classical physics particle/wave nature he thought it was. It turned out to be the taproot of quantum mechanics. The experiment remains as inexplicable now as then. Young showed that when the light goes through two slits, it looks like the familiar pattern waves make when an object is splashed in water or makes a noise in the air. Water and soundwaves are propagation waves of standing water and air molecules transmitting energy by bumping into each other. Waves of this nature interfere with each other when emitted from two sources. At some points, the waves manifest constructive interference where their crests combine to make larger crests, while troughs combine to make deeper troughs. At other points, there is destructive inQQterference where crests and troughs cancel. Noise cancellation devices work by emitting "anti-noise" signals out of phase with environmental noise, so that sound waves cancel by destructive interference. Light makes the same constructive and destructive interference patterns when it goes through two slits, thus bolstering the theory that it travels as a wave.
Then Young blocked one of the slits, expecting the wave behavior to continue, as shown in the middle of the picture. However, wave behavior vanished. Light shot through the single opening lIke a jet of water moving through air. Light behaves like a wave when it goes through two slits and like a jet of particles when it goes through one.

Young's idea that light behaves as a wave was called "prurient" because Isaac Newton's scientific legacy was influential, and Newton had theorized that light traveled as particles. But perhaps light has a dual nature. Maybe it travels as particles through space that generates electromagnetic fields as they move. Perhaps when a light goes through two slits, the electromagnetic fields interfere with each other like waves of water, but when the light goes through only one slit, the fields don't interfere, and the photons shoot across space as particles.

The light was eventually shown to move through space in precisely that way, as photons that generate oscillating waves of electrical and magnetic fields as they move. This caused the light to behave like a "classical" wave of water or sound that interferes with itself when it goes through slits. This type of interference is known as diffraction. However, the slits must be microscopic to make the tiny electromagnetic waves bend around objects and cause diffraction interference. This effect is not visible with large objects. If you place a physical barrier between yourself and the sun, you don't see the light bending into the shadow. However, the noise from a passing bird that "chirps" behind the barrier isn't blocked because noise is a kinetic wave through a medium of air that bends around surfaces. The interference of light that Young saw through the double slits was a different phenomenon. To assume it was caused by "waves," bending around the edges of the slits would be as fallacious as assuming that all waves of water caused by wind, tides, and tsunamis originate in the same way.

In 1983, it became possible to fire photons through the slits one at a time. The one-off photons also generated interference. How could a photon interfere with itself ? That could only happen if photons were traveling as spread-out "waves" much larger than the electromagnetic waves we already knew about. But if each photon is traveling as a large spread out "classical" wave, we'd expect most of it to impact the opaque barrier around the slits. At the same time, a smaller part of it would pass through to impact the measurement screen to make an interference pattern from little bits and pieces of each photon.

However, when photons are fired through the double slits one at a time, it is an all-or-nothing event. Photons are always detected as whole units that land at one point. Electronic devices and our eyes see them that way. Either the entire photon gets through the slits and lands in one whole piece on the other side, or none of it does. The barrier stops most photons. Those that get through the slits build up interference patterns one-by-one on the measurement screen like they do when trillions go through the slit simultaneously in a light beam.

A photon interfering with itself is, therefore, inexplicable by classical wave mechanics. However, photons are massless particles that travel at the constant speed of light without experiencing the passing of time. Maybe that's what allows a photon to interfere with itself. Then it was shown that the same thing happens with electrons, atoms, and molecules made up of many atoms. These are particles with mass that travel at less than the speed of light and therefore experience the passing of time, so it is not even theoretically possible that the same particle might be interfering with its past or future incarnations. Like photons, these particles of mass create interference patterns when going through two slits, while creating jets of particles going through one. It seems that all objects do this, up to some arbitrary, and still to be determined, size. Thus, these mystery waves seem to apply to everything.

It must be that particles move across space as a "cloud" spread out over a wide area. When a particle-cloud encounters a barrier with two slits, it either materializes as an impact point on the barrier, or it passes through both slits as two clouds that spread out beyond the barrier and interfere with each other, thereby warping each particle's impact point into a pattern of interference bands that manifest themselves after many particles are fired through the slits one by one:

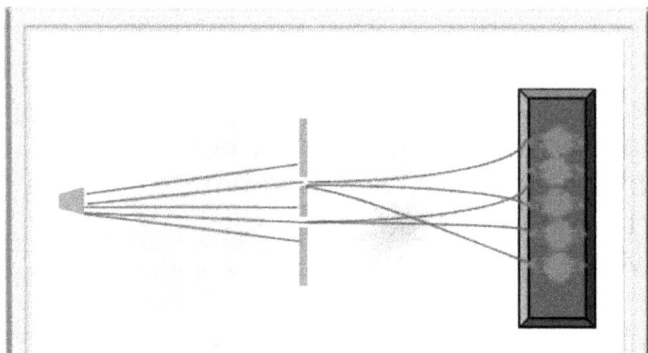

When a particle-cloud encounters a barrier with only one slit open, the particle either materializes as an impact point on the barrier, or it goes through the slit in a straight line without interference and materializes as an impact point on the measurement screen.

When many particles are fired through the single slit one-by-one, they land close together, making a "clump pattern":

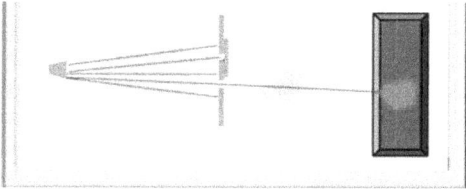

It gets more intriguing. We don't need to block one of the slits to eliminate the interference pattern and make the particles land in a clump.

We can leave both slits open and eliminate the interference pattern just by identifying which of the slits each particle passes through. We could do this by placing a detector to each slit, but it turns out we only need one detector placed to one of the slits.

If the detector is turned off, as shown in the upper half of the picture below, the particles make an interference pattern.

As soon as the detector is turned on, as shown in the bottom half, the interference pattern changes to a clumped pattern:

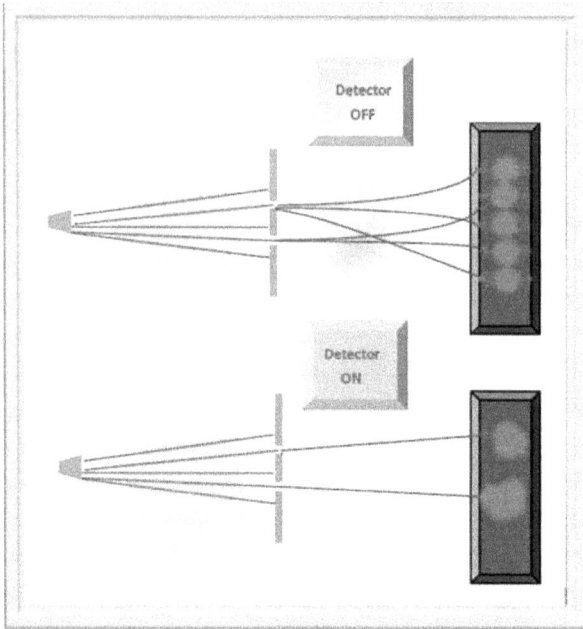

If the electron passes through the slit with the detector, its electric charge imparts information about its location to the detector that reveals the electron's position and causes it to materialize into reality as a straight line particle track that doesn't interfere with itself. An interaction has taken place whereby the electron's electric field has influenced the electrons in the detector, causing it to send a signal to a memory device that records the electron's passing. We might theorize that the detector has also influenced the electron in some way that causes the electron to materialize.

If the electron passes through the slit that doesn't have the detector, then no interaction takes place. Yet the electron also materializes into a particle that doesn't interfere with itself. This is because we can infer that if the electron got to the measurement screen without activating the detector, it had to pass through the slit without the detector.

Thus, it seems that it isn't interaction. Still, rather an information, including inferred information, of which path a particle takes through the slits is what converts it from a cloud into a particle without there being any interaction at all with a detector. Once it becomes a particle, it takes on defined characteristics.

This ability to manipulate a particle by controlling the information we know about it implies that we may have the power to control some aspects of the Universe with our minds by controlling how much information we choose to observe about it. Of course, these tiny particles are so small as to make no difference in themselves. Still, some interpretations of quantum mechanics postulate a quantum chain reaction from the smallest particles to the largest, all the way up to the Universe itself. Because a photon or electron is part of the Universe, changing the state about our knowledge of these small particles affects our state of knowledge of the entire Universe and therefore changes it.

This notion of changing the Universe by gleaning information about one subatomic particle sounds preposterous until we think about how we can initiate stupendous chain reactions in hydrogen bombs. A few

quarts of hydrogen is enough to destroy a large metropolitan area after a fusion chain reaction in one hydrogen atom is ignited. We could, with modest effort, chain enough hydrogen atoms together to blow the surface of the Earth off into space. If we can manipulate something tiny by controlling how much information we know about it, there may be no upward limit to the chain reaction we can manipulate. Because our minds store information, we come to the question of whether they have the power to shape some aspects of the Universe.

Let's Theorize That Our Brains Might Operate On Two Levels:

Level I. Our brains are electromechanical "machines" that store information like computer memory. If this is the highest level our brains are capable of, then consciousness is our mind's illusion to make us think we are making decisions when we aren't.

Level II. Our consciousness is a product of mind, separate from the Universe of matter, energy, space, and time. Perhaps our minds are so separate from the material Universe that they cannot be explained by material processes such as the electrical impulses in our brains that affect its atoms and molecules.

CONCLUSION

Physicists can have it rough when confronted with mysterious problems, not because they have any more issues than they would have if the problems weren't mysterious, but because popular culture tended to convolute an understanding of the mystery. It is no surprise that the media and movies can twist science when science does not specify precisely what it is talking about. It has been speculated that there is a different type of matter in our universe that accounts for certain mysterious observations. Galaxies have been heavier than they should be, stars move faster than they should move, and particles collide unexpectedly. A string of explanations has been provided by research scientists in astronomy. In general, these explanations incorporate the existence of dark matter.

The reason why dark matter is "dark" is that it is invisible to the human eye and every telescope technology built thus far. What that means for research purposes is that dark matter cannot be as easily experimented with, as can particles that fall under the definition of "matter." If you want to test Newton's laws of gravitation, you can take a banana, put it next to Venus, and measure the gravitational force on the banana as a function of whatever variable you want. You can look at the banana's orbital acceleration, for example, but you cannot find yourself a dark banana, put it next to Venus, and understand how it behaves in response to varying physical parameters.[12]

12 No, it does not matter how ripe the banana is.

The universe consists of 27% dark matter, which poses a very interesting question: how is it that over one-fourth of the universe is comprised of something we can't even see? Moreover, how has there been so much experimental support for something we can't work with?

A Dark Matter Galaxy?

The answer is that you don't need to see something to "do science" on it. Two branches of physics have proven very promising areas of experimentation: astrophysics and particle physics. Very recently, a team of researchers led by Peter van Dokkum at the W. M. Keck Observatory in Hawaii discovered an intriguing galaxy named Dragonfly 44. Dragonfly 44 has a very small collection of stars. Van Dokkum's objective was very simple: he knew the stars would be moving at some orbital speed. Based on previous literature, however, he also understood that certain stars in Dragonfly 44 moved far too quickly for what their actual speed should be, according to calculations. The strategy was to evaluate what the orbital speed was supposed to be and measure what the speeds of the stars were. Comparative analysis between these values might reveal interesting (or maybe not!) possibilities.

The expected orbital speed of the stars is not incredibly difficult to calculate. Two theories of gravitation can be used to do so: Newtonian's law of gravitation or Einstein's theory of general relativity. According to the former, the more matter there is in a system, the larger the magnitude of the force on each component. The larger the force, the greater the acceleration, and the greater the change in orbital speed. According to the latter, gravity is related to spacetime curvature, but there is a similar argument: more matter in the spacetime map leads to more curvature, meaning stronger gravitational effects. Both situations postulate that the more matter there is in a given area of space, the greater the change in orbital speed should be of the components (stars) that make up that system. Because there were relatively few stars in Dragonfly 44, it was assumed that the stars could not be moving very fast. However, when they calculated experimental orbital speeds, they ended up with values that were significantly greater than what they should have been. It doesn't matter which team you're on, Newton or Einstein because you're ultimately going to come to the same conclusion: if these stars were moving faster than they should be moving, there has to be more mass in the system. Precisely, the mass that scientists before Van Dokkum could not detect.

It fit the dark matter description quite nicely. If Dragonfly 44 were composed of dark matter, it would have a mass that we can't measure. That extra mass might contribute to gravitational effects experienced by the stars, which is why they move with such fast orbital velocities. The next objective was to dig a bit deeper: if it could be calculated how much mass was necessary to force the stars to move at their speed, it could be known how much dark matter is present in this galaxy. It turned out. Dragonfly 44 is composed of dark matter by nearly 99%. Van Dokkum and his team found a dark matter galaxy.

There is a particle accelerator at the CERN laboratory that collides large hadrons or a specific class of particles. Despite nearly a decade of debate about a unique name for the collider, participating physicists settled on the Large Hadron Collider (LHC). In the LHC, particles can be crashed into each other to produce collisions. When two particles collide, two variables are important: the energy and momentum changes of the collision. Generally, in ideally isolated systems, both energy and momentum of collisions are conserved. That means if you send two bananas crashing into each other and watch the two bananas stick together after colliding, the energy of one banana added to the energy of the other banana will give you the total energy of the stuck-together bananas at the end. The same applies for momentum. Yet, that's not always the case. If baryons and fermions, members of other classes of particles, are sent moving towards each other, momentum and energy conservation is generally not observed. Energy and momentum tend to both be lost. However, if you force the baryons and fermions to collide ideally, meaning you deprive the system of everything but these two particles, you should get momentum conservation. That's exactly what has happened at the LHC: baryons and fermions have been put in isolated systems with the momentum before and after the collision both measured. What should have been no difference became a very obvious difference: momentum was lost.

It was, therefore, postulated that dark matter particles were produced in these collisions. Such dark matter particles carried away some momentum to produce the change that was measured. Because dark matter can slip from detection, necessary measurements to verify these postulations could not be made directly. When the momentum of these supposed dark matter particles was measured and compared to the momentum change of these particle collisions, the data were exceptionally telling.

www.ingramcontent.com/pod-product-compliance
Lightning Source LLC
Chambersburg PA
CBHW071716210326
41597CB00017B/2508